21st Century
Basic Skills
Library

BAR GRAPHS

Candy Colors

by Sherra G. Edgar

Cherry Lake Publishing • Ann Arbor, Michigan

2

Published in the United States of America by Cherry Lake Publishing
Ann Arbor, Michigan
www.cherrylakepublishing.com

Consultants: Janice Bradley, PhD, Mathematically Connected
Communities, New Mexico State University; Marla Conn, Read-Ability
Editorial direction and book production: Red Line Editorial

Photo Credits: Nate Allred/Shutterstock Images, cover, 1; Tatyana
Vychegzhanina/Shutterstock Images, 4; Shutterstock Images, 6, 8,
16; Michael Jung/Shutterstock Images, 10; Mat Hayward/Shutterstock
Images, 14; Diego Cervo/Shutterstock Images, 18; Jiri Hera/Shutterstock
Images, 20

Library of Congress Cataloging-in-Publication Data
Edgar, Sherra G.
  Bar graphs / Sherra G. Edgar.
      pages cm. -- (Let's make graphs)
  Audience: K to grade 3.
  Includes bibliographical references and index.
  ISBN 978-1-62431-390-5 (hardcover) -- ISBN 978-1-62431-466-7
(paperback) -- ISBN 978-1-62431-428-5 (pdf) -- ISBN 978-1-62431-504-6
(ebook)
  1. Graphic methods--Juvenile literature. I. Title.

  QA90.E34 2013
  518'.23--dc23

                          2013004937

Cherry Lake Publishing would like to acknowledge the work of The
Partnership for 21st Century Skills. Please visit www.p21.org for more
information.

Printed in the United States of America
Corporate Graphics Inc.
July 2013
CLFA11

TABLE OF CONTENTS

What Is a Bar Graph?

Anna wants to count what kinds of fruit she has. She can use a **graph**. Graphs show **data**.

Fruits

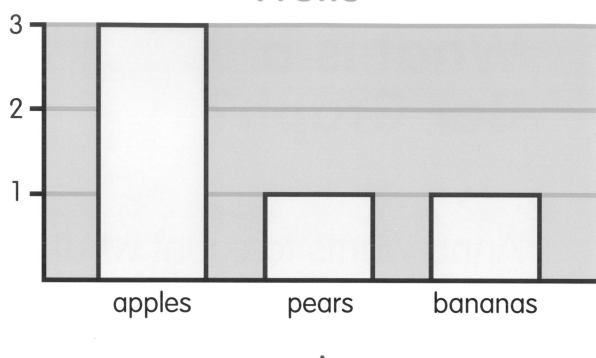

Anna made a **bar graph**. On a bar graph, **bars** stand for **amounts**.

Game Pieces

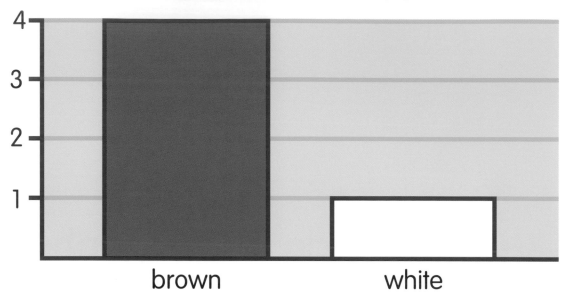

brown   white

Longer bars mean bigger amounts. Shorter bars mean smaller amounts. These bars help us **compare** data.

Making a Bar Graph

Jed wants to count people's favorite toys. He calls his graph "Best Toys."

Best Toys

10
9
8
7
6
5
4
3
2
1

Jed made a big **L**.
He numbered 1 to 10
up the side of the **L**.

Best Toys

10	
9	
8	
7	
6	
5	
4	
3	
2	
1	

games   books   bikes

He wrote **labels** under the
L: games, books, and bikes.
These are the things Jed
will count.

Best Toys

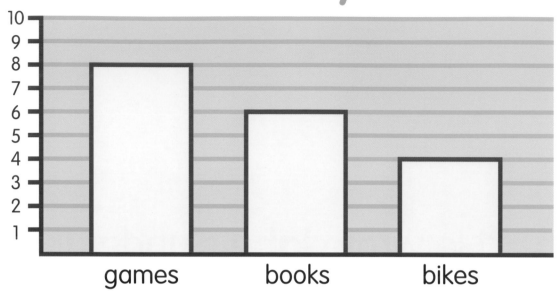

Jed made a bar for each toy.
Eight friends like games best.
Six friends like books best.
Four friends like bikes best.

Best Toys

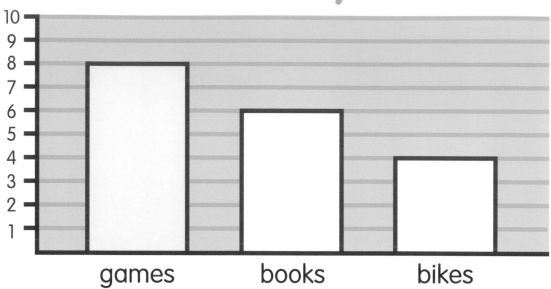

What does Jed's graph show? You can see Jed's friends like games best.

You Try It!

Make a bar graph with blocks. Sort the blocks by color. Count the blocks. Make the graph. Graphs are fun!

Find Out More

BOOK

Bader, Bonnie. *Graphs*. New York: Grosset and Dunlap, 2003.

WEB SITE

NCES Kids' Zone: Create a Graph
http://nces.ed.gov/nceskids/createagraph/
Use this online tool to make your own graphs.

Glossary

amounts (uh-MOUNTS) how many or how much there is of something

bars (BAHRS) solid rectangles that stand for numbers on a bar graph

bar graph (BAHR GRAF) a graph that uses bars to stand for numbers

compare (kuhm-PAIR) to show how things are alike

data (DEY-tah) amounts from a graph

graph (GRAF) a picture that compares two or more amounts

labels (LEY-buhls) names

Home and School Connection

Use this list of words from the book to help your child become a better reader. Word games and writing activities can help beginning readers reinforce literacy skills.

a	data	kinds	sort
amounts	does	labels	stand
and	each	like	the
Anna	eight	longer	these
apple	favorite	made	things
are	for	make	to
banana	four	making	toy
bar	friends	mean	toys
best	fruit	numbered	try
big	fruits	of	under
bigger	fun	on	up
bikes	game	pear	us
blocks	games	people's	use
books	graph	pieces	wants
brown	has	see	what
by	he	she	white
calls	help	shorter	will
can	his	show	with
color	is	side	wrote
compare	it	six	you
count	Jed	smaller	

Index

About the Author

Sherra G. Edgar is a former primary school teacher who now writes books for children. She also writes a blog for women. She lives in Texas with her husband and son. She loves reading, writing, and spending time with friends and family.